Software Project Reviews

Richa Yamini Goel

SOFTWARE PROJECT REVIEWS

CONTENTS

1. CONTENTS
2. INTRODUCTION
    Static vs Dynamic Testing
    Review Principles
    What to Reviews?
    Why review?
3. REVIEW TECHNIQUES
    Walkthrough
    Inspection
    Informal Reviews
4. REVIEW PROCESS
    Who should perform reviews?
    How to perform Reviews?
    How review should be done?
    What need to be checked:
5. ROLES IN REVIEW
6. AGILE REVIEWS
7. REVIEW METRICS
8. KEY GUIDELINES

SOFTWARE PROJECT REVIEWS

Preface

This book is written with the focus of putting in all the information related to project reviews from the purview of project management. A process of reviewing the documents, to provide a better quality product to the end-user, is a necessity in today's competitive world.

Reviews can result in saving cost and time in a long run. But still in a lot of software projects, reviews remain the neglected phase of project management.

The concepts, challenges, and ideas mentioned in this book are taken from the practical implementation and my personal experiences in the industry.

I also want to extend my thanks to my genuinely nice friends and colleagues, Ruchi Kalra, and Anamika Chaudhary, who contributed their experiences and ideas and help me in the editing of the book.

I also want to thank my publishers for motivating me to write down my experiences and communicate them to others using various mediums.

SOFTWARE PROJECT REVIEWS

SOFTWARE PROJECT REVIEWS

Introduction

Test Reviews are a very crucial part of the Software Development Life Cycle (SDLC). The reviews are embedded in all phases of the Software Development Cycle. The test reviews are part of Static type of testing, and an important part of Quality Assurance.

A **software review** is a systematic process or meeting during which a software product and/ or its components are examined by one or more individuals, users, computers, user representatives, or other interested parties for comment or approval, to validate the quality, functionality and other vital features and components of the software.

It is usually performed manually and is used to verify various documents like requirements, system designs, code, test plans, and test cases, against a standard or a practice.

The concept of Test reviews is dated back to the 1950s, when the concept of Total Quality Control was introduced in 1951 and, the program testing and debugging were explored in 1957 by Mr. Charles L Baker. The concept of Reviews is further explored in 1959 using the PERT concept, i.e.; Program Evaluation and Review Technique.

In 1986, the V Model was introduced which demonstrates the relationships between each phase of the development life cycle and its associated phase of testing, in the form of static as well as dynamic testing. This gives way to the concept of reviews which then explored and improved with time. The importance of the reviews is hence realized.

The main objective behind doing the software reviews is to improve the productivity of the development team, to make the testing process time and cost-effective, to make the final software with fewer defects, and to eliminate the inadequacies. The Review is a process of looking over a document/piece of code for the purpose of evaluation, for examining the quality or for approval.

Static vs Dynamic Testing

Static testing is a way of testing that is done by not running the software but looking and going through at its components like code, requirements, help documents, user manuals, etc. They are not executed, but tested with the set of some tools and processes, i.e.; inspecting without executing. It provides a powerful way to improve the quality and productivity of software development. This process is also known as dry run testing or a process of verification.

Dynamic Testing is basically a way of testing when the actual execution is done on the software code as a technique to detect defects and to determine the quality attributes of the code. With dynamic testing methods, the software is executed using a set of inputs, and its output is then compared to the expected results. This process can be done manually or via automation tools. This process comes under validation.

Static Reviews provides a powerful way to check and improve the quality and productivity of software products and to recognize and fix their own defects early in the software development process in the form of code reviews, requirement reviews, test object reviews, etc. Nowadays, all software organizations are conducting reviews in all major aspects of their work including requirements, design, implementation, testing, and maintenance.

With the due course of time, the process of reviews has evolved a lot and is used for reducing the rework. Test reviews have been a very important part of any SDLC but have been ignored until now. The reviews are done in a project but not with the right intention. The reviews can play a driver for saving cost and we want to explore the potentials of reviews within the SDLC.

Reviews are an integral part of the project lifecycle and a critical process. In the review process, the reviewing person checks to see if the changes made to the code, data, or documents are correct or not. In other words, it is an analysis that has been conducted at a fixed time to evaluate how far the stated goals have been achieved.

Review Principles

Before starting off with the test reviews, it is important to understand the principles to focus upon:

- **Adherence to standards:** It is really important to adhere to the standards on the basis of which the review needs to be done. The standard or the basis should be clear and should be shared with both reviewer and author.

- **Partial Knowledge of reviewer:** The reviewer should be well aware of the requirements and should have access to all supporting documents. The partial or no knowledge of the requirements will defeat the purpose of the review with fruitless output.

- **Reviewer Attitude:** It is important to focus on the review object instead of focusing on the fault-finding. The purpose of the review is to improve it.

- **Review the product not the producer:** Ensure the purpose of the review is to judge the review object and not the author of the object.

Static test reviews is a disciplined approach to evaluate whether a software product fulfills the requirements or conditions imposed on them in the form of standards and guidelines, or in other words, are we doing the job right?

The test reviews can be performed using various techniques like walkthrough, formal inspection, informal reviews, and review of each software product. The primary practice is to ensure that it is done systematically by reading the contents or components of a software product with the intention of detecting gaps. It is the process of reviewing /inspecting deliverables throughout the life cycle.

Another significance of performing reviews is that all the team members get to know about the requirements and progress of the project and sometimes the diversity of thoughts may result in excellent suggestions to improve the documents. These documents are directly examined by people and discrepancies are sorted out at the same time.

The reviews with time have proven its merit in all software projects, hence the Organizations are allocating time in the project plans for reviews. Effective reviews not only save money and time in a project but also improves the quality of the product and maturity of the project members.

Review Components

What to Reviews?

It is really important to understand what all components within the SDLC (Software Development Life Cycle) can be reviewed. There are multiple documents that are created during the creation of software. All such documents can be reviewed to improve the quality; although due to time constraints it is difficult to review all of them. Hence, selective important documents are reviewed, like:

- Requirement Specifications
- Functional Specifications
- Design Specifications
- Code
- User's Guide
- Test Plan
- Test Specification (Test Cases)
- Test Script

Each SDLC phase is a translation, or output from the previous phase and it creates a work product that can be tested to see how successful the translation is. As stated by the top researchers, at the early stage of the development lifecycle, is the majority of all defects i.e. around 50% of all defects come only from user requirements, which is found in early stages, can save a lot of money and time.

A review process should prevent this defect migration if done on time. The cost of defect migration to the next stage is much higher than finding a defect in the phase where it was introduced. At the next stage, it can cost an order of magnitude more and an order of magnitude more again at the stage after that. The cost is maximized if the error is detected after the product is shipped to the customer and minimized if it is detected in the phase where it was introduced.

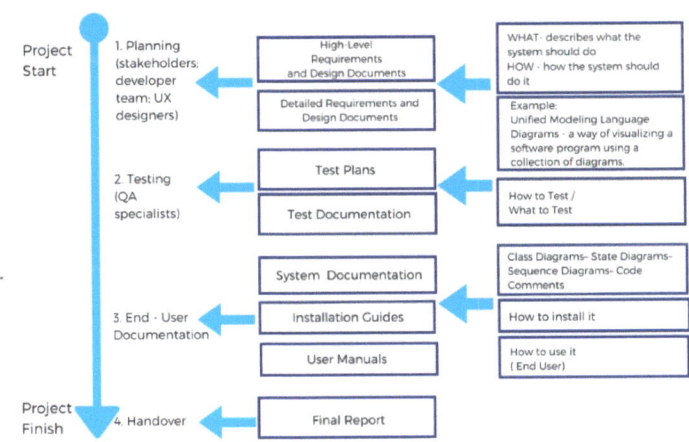

PROJECT DOCUMENTATION IN SDLC PROCESS

There are various documents created in each stage of the development lifecycle. Here is a high-level overview of a few of the major documents created in any software project.

Process Flow →

Requirement Gathering	Statement of Work (SOW)		Business Requirements	User Requirements	Functional Requirements	Non Functional Requirements	Requirement Traceability matrix	
Design & analysis			High Level Design	Use Case Document	Low Level Design	Query analysis		UI document
Development/ Coding	Coding Guidelines		Code					Unit Test Plan & Test Cases
Testing		Project Plan	Test Plan	Integration Test Cases	Regression Test Cases	Automation Test scripts		Defect Report
Delivery			Deployment Plan	User Manual	Delivery Notes	Installation Guidelines		
Maintenance & Acceptance			UAT Test Cases	UAT Defect Report	Change Requests & New Requirements			

The blue boxes are the documents which are usually involved in the review process.

DOCUMENTS MAPPED WITH SDLC PROCESS

Let us have a look at some of the documents which can be reviewed in each stage.

- *Requirement Specifications:*

The purpose of the requirements phase is to ensure that the users' requirements are properly understood before translating them into a core design. In this phase, the purpose, scope, and performance of the required deliverable are defined. The review of the completeness of these requirements is done. The focus should be on missing requirements and requirements gap.

- *Functional Specifications:*

The functional design is the process of translating user requirements into the set of external interfaces. In this phase, the scope, characteristics, and performance criteria of the system in terms of hardware and software, that meet the user requirements, are defined. The review of the specifications is done with the perspective of development, testing, database, usability and installation purpose.

- *Design Specifications:*

The internal design is the process of translating the system requirements into a detailed set of data structures, data flows, and algorithms. In this phase, the most appropriate physical solution, positioning against existing architecture and applications to meet the agreed system requirements are specified. The review is performed against the internal design (Architectural Design, Module Design Specifications, Database Schema, etc.)

- *Code*

Coding is the process of translating the internal design specification into a specific set of coding languages. The code is the most critical part of the software, therefore, it requires a thorough review to avoid major defects in the future.

The review process for code should focus on:
- Data declaration and reference errors
- Computation errors
- Comparison of errors
- Control flow errors
- Interface errors
- Input/ Output errors
- ...

- *User's Guide*

Every software product that is delivered to a customer consists of both the executable code and the user manuals. A product's documentation is no less important than its code because its quality is a significant factor in the success or failure of a product. From the point of view of the user, if the manual says to do something, and the user follows these instructions and it doesn't work, then that is a defect in the manual.

It is important to ensure here that the manual is reviewed to ensure the correctness of the steps and guidelines mentioned.

- *Test Plan & Scenarios*

The testing process is as important as the development of the software. The quality of the software is dependent on the testing team. Hence, it becomes really significant to ensure the review of the testing artifacts like test plan and test scenarios.

The focus of the review should be the selection of:
- Test scenarios
- Test data
- Test case steps
- Test results
- Test Criteria
- Test Pre-requisite

The main reason for reviewing the test cases is to increase test coverage and therefore product quality. A review of the test case functionality is done as per the functional document.

Less number of defects and better quality of defects are likely to encounter if the test artifacts are properly reviewed. The document errors take more time in fixing, as it involves discussions, requirement approvals, and verifications from all stakeholders.

Why review?

It is important to understand the purpose of reviews before we begin. The reviews help us in various ways, some of which are listed below:

- *Defect Variants*
  The type of defects detected via the review techniques are deviations from standards, missing requirements, design defects, non-maintainable code and inconsistent interface specifications, many more.

- *Cost-effective*
  It is simpler and cheaper to correct faults early in the life cycle because:
    - Fewer people involved
    - Less travel needed
    - Less time required for maintenance
    - Less debugging needed

- *Problem prevention*
  Reviews help to find the faults in the code before the software is implemented and reduce the risk of misunderstanding the requirements. Early reviews help is identifying whether the product becomes testable or not.

- *Involve the project team early*
  Project members will bet a common view of the project when involved in reviews early in the lifecycle. The team gets a better understanding of the requirements.

- *Effective learning*
  When reviewing you get a better knowledge about both the product and the project, and this helps in evolving the requirements for better output for the end-user.

Review Techniques

Reviews are essential for improving the quality of the work product. It also gives the author a learning curve and confidence.

There are multiple review techniques used depending upon the type of document to review, time available for review, available resources, severity, and criticality of requirements. The different review techniques are:

- Walkthrough
- Inspection/ Formal Reviews
- Informal Review

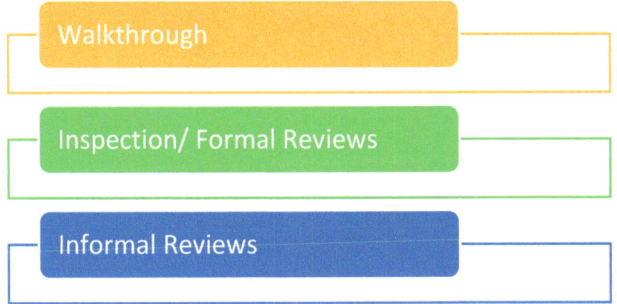

TYPES OF REVIEWS

Walkthrough

A walkthrough is a kind of step-by-step review process, where the document, usually, a piece of code, is reviewed line by line. A walkthrough can be a formal or an informal (non-procedural) activity in the Verification. It is a structured process where multiple reviewers can participate at the same time.

Typically, this type of review is done to perform code reviews, although no advance preparation is required by the participants. This is also known as Dry Runs.

A walkthrough is conducted by the author of the 'document under review' who takes the participants through the document, line by line, and through his or her thought processes, to achieve a common understanding and to gather feedback. A company that has regular walkthroughs may have made their own rules for walkthroughs.

A walkthrough is a group activity, and the main purpose of the same is to train the team and/or imparting knowledge transition.

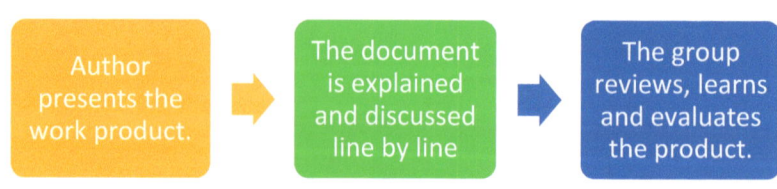

WALKTHROUGH PROCESS

The disadvantage of a walkthrough is that a review tends to be less objective when the author is the producer, and sometimes time-consuming. The author knows the subject well and has to be prepared for questions.

Often the purpose is to achieve consensus with the listener or peer group. This is also useful when the code needs to be explained to all the peers, or a large group.

Inspection

Inspection is a Team activity and is a formal activity. An inspection is a group review quality improvement process for written materials.

The formal reviews are driven by defined phases and the following properties:

- *Planning*
  Formal reviews are time-consuming. To get the best output, time for the reviews should be included in the project plan. Each review should be rigorously planned.

- *Documented*
  The review process has to be documented. This way the review tasks should be possible to run in the same way from one time to another. The

more rules are specified for the review, the more formal it is.

- *Thorough*
  The document which is being reviewed is checked against other documents to verify that it is correctly derived from specifications and standards.

- *Focused on a certain purpose*
  The reviewer reviews the document in advance and from a specific view and against the defined agenda.

Formal Reviews contain roles like Inspector, Moderator, Reader, Recorder and Author. The document under inspection is prepared and checked thoroughly by the reviewers before the meeting, comparing the work product with its sources and other referenced documents, and using rules and checklists. In the inspection meeting the defects found are logged.

Inspection or a formal review requires a formal entry, usually in the form of a meeting invite, and a formal exit, usually a document of findings. The formal review comprises of:

- *Defined adaptable formal process*

Defined rules and checklists govern the work in an inspection process. The process, formats, and the procedure are already defined by the Organization.

Software Project Reviews

- *Defined Roles*

Each participant in the review can have one or more roles in the process (moderator, author, reviewer, manager, review manager) and corresponding responsibilities.

- *Defined Deliverables*

The inspected document is the most important delivery but there are other types of deliverables as well like defect list, suggestions, and open-ended feedback.

The inspection can further be divided into different sections – Online Inspection, Offline inspection, Audit Reviews and Management Reviews.

Online Inspection

In online formal reviews or inspections, the entire group, i.e.; identified reviewer(s) and the reviewee, sits together with printouts, projectors, etc. online and discusses the gaps and issues. They either use video conferencing or sit in a meeting room and do the review together.

All the participants involved are fully responsible for the quality of the review and for the quality of the information. Active and open participation of everyone in the review group, is required.

This kind of inspection is typically used for requirements and high-level design reviews using checklists, rules, metrics, etc., and majorly involves two aspects:

- product (document itself) improvement and
- process improvement (of both document production and inspection).

Offline Inspection

In a formal offline review, the work product is reviewed in the absence of the Author. The involved participants are only the reviewer(s) and the reviewee. The reviewer(s) do individual review and send their comments and issues using an email program, or log the same in the defect management system.

For Formal Offline review,
1. The author gives the work product to the reviewer(s).
2. The author should send the item that should be reviewed along with related items, to the reviewer.
3. The reviewer(s) review the work product and prepare a Review Report.

Offline inspections save time while ensuring the defects are captured similar to the formal online inspection process. Although there is a disadvantage in this process as well. There is very less, or no scope for discussion in the offline review process.

The offline reviews are not recommended for the complex or critical functionalities.

Software Management Review
Software Management Review evaluates the work status. On the basis of these reviews, decisions regarding downstream activities are taken.

Management Reviews necessarily involve top management as the reviewer in the project, who review the progress of the project. In Management reviews, no detailed discussions happen, and only high-level updates are discussed.

The purpose of such reviews is to let the management know the current status of the project, and inform them about any kind of hurdles encountered in the execution of the project.

Software Audit Review:
Typically known as IT Audit, or Asset Audits, Software Audit Review is a type of external review in which one or more critics, who are not a part of the internal team (within or outside the organization), organize an independent inspection of the software product and its processes to assess their compliance with stated specifications and standards. This is done by managerial level people.

The purpose of such reviews is to understand and analyze the security flaws, open ports, unauthorized accesses, unlicensed software etc.

Informal Reviews

An informal review is a process where usually one reviewer conducts the informal review, alone. The results of the same are not recorded. Usually informal review is done to make the product ready for formal reviews.

In this review, the reviewer could act as the reviser along with the author. Typically used for screen designs, small routines that do not require a formal inspection or a walkthrough.

An informal review (also known as buddy checks) is better than no review, provided it is performed by someone other than the author and its objective is to detect defects. However, simply giving a document to someone else and asking them to look at it closely will turn up defects we might never find on our own.

It is quite similar to the situation where the teacher has given you an assignment. You went to your friend to get your assignment verified offline before the teacher checks it and grades it. This process is similar to an Informal Review. But when the teacher formally checks your assignment and grades it, it is a formal review process.

The informal reviews are driven by defined phases and the following properties:
- Undocumented
- Fast
- Few defined procedures
- Useful to check that the author is on track

- *Undocumented*

It is an undocumented review process, where no plan and comments are documented. The whole process is completely informal and in the form of a favour.

- *Fast*

The informal reviews are faster in nature as compares to formal reviews. Because it is undocumented, hence the whole document is reviewed at a glance i.e.; many pages per hour, or many lines of code in a minute.

- *Few defined procedures*

Since the informal review is undocumented, the defined procedures for how the review shall be carried out is usually few or none. The normal situation is that no entry criteria are enforced, no roles are defined, no required skills are required and the agenda is informal.

- *Useful to check that the author is on track*

It gives feedback to the author about the document and checks that it proceeds in the right way. It may help the author to gather information, find the requirements and evaluate implementation ideas, when this cannot be achieved by other means, such as the study of source documents, requirements modeling, developers' meetings, etc.

Summary

Depending upon the type of document, and its severity the review technique can be chosen. The review technique may include a combination of two or more techniques.

For example, a critical and complex functionality may first get informally reviewed and then formally reviewed, for getting the best output. It may also be possible that after the formal review or walkthrough is done, the recommended changes or defects are verified informally.

It does not matter which particular technique is used. But the focus should be on performing the review in the correct manner and with the right intentions.

The difference among all 3 techniques is stated as a summary below:

	Walkthrough	**Informal Reviews**	**Formal Reviews**
Pre-requisite	The meeting is planned and the copy of the document to be reviewed should be distributed to the group of reviewers.	The document to be review should be given to the reviewer.	Meeting requests need to be raised and the document under review is shared with the reviewers in advance along with the review guidelines.
Entry Criteria	When all planned review members arrive.	Whenever the document to review is received.	The coordinator starts the review meeting.
Team Involved	Author(s) and a group of reviewers	an Author and a Reviewer	Author(s), Reviewer(s), Moderator and Recorder
Findings	Findings are suggested on the spot and recorded by the author.	Recommendations are given by the reviewer.	Recorder notes down the recommendations and findings and a formal report is prepared.
Exit Criteria	The author updates the document as per the feedbacks received.	Recommendations are received by the author.	The recommendations and findings are incorporated and the same is verified by the reviewer.

Review Process

Who should perform reviews?

The review is an essential part of project management. Hence, the reviews should be done by an individual who can be accountable for the gaps in the document. The reviews should be done by the individual who:

- Has the knowledge of the document under review, and experience in similar projects.
- Can devote the time for review.
- Can take the accountability of the review defects.
- Can document/ recommend the review defects to the author.
- Has Technical skills, and listening skills.
- Should be impartial, and practical.
- Should be open for change.

The point to remember is that the output of a review activity is fruitful only if it is done by someone other than the author.

How to perform Reviews?

The review process is very simple but can be time-consuming. There are some key points which we need to keep in mind while performing the reviews, and to manage the project time.

HOW TO REVIEW

- *Split big documents into smaller parts*

Sometimes a document under review is quite huge in size, and/ or moderately complex in nature. Hence, it is more efficient to split the document into smaller logical parts. The reviewer can then manage to review the whole document with the help of other reviewers by dividing the sections among themselves. The reviews would be more efficient and quick.

- *Samples*

This is one way to reduce the effort of reviewing a large document. E.g. you can pick out some of the most important pages or pick out every tenth page of the document and call for a formal review. Every reviewer is reviewing the same pages and analyze the kind of defects and advise to fix the similar changes in the complete document.

- *Let different types of project members review*

Different people find different things because they have different views. So, the review team should include the project members/ reviewers with different technical skills, like coding, testing, database, networking, etc. Such kind of mixed reviewers will bring the best output as they will review the documents from their needs and perspective.

- *Select a suitable review type*

It is equally important to identify the suitable review type which can do justice to the time devoted by different team members in the review activity.

- *Review guidelines should be in place*

The review guidelines should be drafted and communicated to all reviewers and reviewees before the process starts. Review materials, along with the checklists and formats, if any, must be distributed in advance of the review meeting. The meeting duration, entry, and exit criteria must be communicated.

How review should be done?

The reviewing process, whether formal or informal, highly depends on the complexity of the document to review, the size of the document and the entry and exit criteria. The review has to be a transparent process. Let us look at them, one by one.

Walkthrough Process

The entry criteria for the walkthrough starts with the arrival of the members of the review team. The purpose is to provide a proper walkthrough to each member, to make them understand each line of the document under review. The exit criteria are whenever the author of the document is done by explaining the last line of the document.

The output of walkthroughs is in the form of suggestions from the reviewers. Incorporating those suggestions or not, is the author's decision.

Informal Review Process

The entry criteria for the informal reviews is whenever the document is handed over to the reviewer by the author. And the exit criteria for the same is when the reviewer provides the feedback in either written or verbal format to the reviewee.

The gaps/ recommendations found in an informal review are not known as defects as these are the suggestions provided by the reviewer. The closure of these recommendations is up to the reviewee. If the reviewee does not want to incorporate the suggested changes, then also the informal review activity is considered as closed.

The informal reviews are done one on one, hence no recorder or time limit is considered in the same. In many projects, no separate time is allocated for informal reviews.

Inspection Process

Inspections are the formal review process and are expected to follow a particular life cycle. The entry criteria for the inspection is the meeting request or planning process. Before the review meeting entry criteria are enforced on the inspected object to see that meets minimum quality standards required for the inspection to be worthwhile, e.g. that spell checker has been used. Every organization has its internal formal process, but the important phases are as below:

1. Planning
2. Preparation (individual)
3. Review meeting
4. Rework
5. Follow-up
6. Metrics and Process improvements

1. *Planning*

The inspection leader often makes the review plan. The review plan includes when what and how to perform the review activity, who should perform the review and what will be the roles etc. The meeting is scheduled and the plan is sent to all the participants. Sometimes the copy of the document is also shared to save the study time in the meeting.

The review process for a particular review begins with a 'request for review' by the author to the moderator (or inspection leader). A moderator is often assigned to take care of the scheduling (dates, time, place, and invitation) of the review. The project planning needs to allow time for review and rework activities.

2. *Preparation (individual)*

This phase requires the self-study of the document. The reviewers are expected to identify the gaps and concerns to discuss in the review meeting.

In this phase, the reviewers work individually on the document under review using the related documents, procedures, rules and checklists, requirements provided. They identify defects, questions, and comments, according to their understanding of the document and role. Spelling mistakes are recorded on the document under review but not mentioned during the meeting. The annotated document will be given to the author at the end of the logging meeting. Using checklists during this phase can make reviews more effective and efficient.

3. *Review meeting*

At the start of the review meeting entry criteria for the participants are enforced, i.e. have the participants performed the expected tasks. The coordinator/ moderator informs about the agenda and the time frames.

The complete document is reviewed line by line, and page by page. The queries and questions are discussed on a high level

Gaps, concerns and problems found during the review are logged but should not be examined in detail (discussions should be taken afterward to reduce time consumption). The reviewee clears as many concerns and queries during the meeting.

The moderator ensures that the meeting is not diverting from the topic and the time management is maintained. The outcome of the discussions is documented for future reference. The meeting notes are prepared and shared with all participants with action items and expected closure dates.

At the end of the meeting, a decision on the document under review has to be made by the participants, sometimes based on formal exit criteria. The most important exit criterion is the average number of critical and major defects found per page. If the number of defects found per page exceeds a certain level, the document must be reviewed again, after it has been reworked.

4. *Rework*

Based on the defects detected and improvements suggested in the review meeting, the author improves the document under review. In this phase the author would be doing all the rework to ensure that defects detected should be fixed and corrections should be properly implied.

As per the suggestions and gaps suggested, the author will update the document accordingly and resubmit the same to the review team using the informal review process.

5. *Follow-up*

The reviewers are supposed to check that the corrections are made and the statistics are logged (how many faults found, how much time spent). This is done by the inspection leader.

After the rework, the moderator should ensure that satisfactory actions have been taken on all logged defects, improvement suggestions, and change requests.

6. *Metrics & Process Improvements*

Metrics help in improving the process and updating the checklists for a smooth transition of the process. A walkthrough is recorded and the results are then used for process improvements or delivery decisions.

In metrics, the type of defects reported, the count of defects, turnaround time etc. are recorded and the data is used for process improvement. The data analysis is done to understand the gaps encountered.

What needs to be checked:

It is very important to understand what things to be checked when a review is done. Reviewers often divert from the purpose of the process and move towards the personal vendetta. Here are a few important areas listed which need to be kept in mind when a review, whether formal or informal, is performed.

- **Coverage of Topics**
  It is important to check whether all essential topics/ requirements are covered in the document. If covered, the completeness of the requirement/ topic needs to be checked along with the details, assumptions, facts, examples etc.

- **Correctness**
  Whatever information or instructions are mentioned in the document need to be checked for accurateness. There should not be any contradictions. In case required, further details can be provided in the comments.

- **Utility Value**
  The first thing to review in a test case or in a document is its utility value. It is important to check whether the lines of code or the test case is required and is not repetitive. Unnecessary lines should be eliminated.

- **Objective**
  The objective of creating a document should be clear. While creating the document, whether requirement, coding, or testing document, should be clearly outlined and should stick to the objective behind the creation of the document.

- **Pre-Requisites**
    It is important to understand whether there is any pre-requisite required to understand the document under review. List down the reference documents and the pre-requisite knowledge required for the review of the document. Do not assume.

- **Clarity and Details**
    The document should be written with clear instructions and comments. If the document is stating the requirements, then the requirements have to be in the very detail, or if the document is a test case, the steps to reach the screen should be easy to understand. In the case of a coding document, the comment entries should explain what the below code is going to do.

- **Language:**
    The language used in the document should be simple and unambiguous. There should be no scope for interpreting the document in any other manner except the desired one. Do not try to use complex words. The purpose is to make the document easily understand, and not to show off the vocabulary.

- **Output:**
A good document should clearly mention the expected output. It is important for the reader to easily realize the conclusion or the outcome from the document.

- **Clean-up:**
If the document has changed the initial state of the system, then it is important to add the steps to bring the system back to its initial state. In the case of databases, test cases, installations, it is important to add the cleanup section. In case of coding, the steps to flush the cache, clearing the open threads and closing the database connections should be added. In the case of requirements, the purpose of change has to be clearly identified and mentioned in detail.

Similarly, the other aspects of the documents should be reviewed depending upon the type of the document. The purpose should always be to improve the document, by adding something productive to it.

Roles in Review

Who should review it?

In the software development lifecycle (SDLC), there are multiple documents that are required to be reviewed. The reviewer(s) for all these documents can be different roles in the hierarchy. Hence, it is important to select the right reviewer for the document to get the appropriate output.

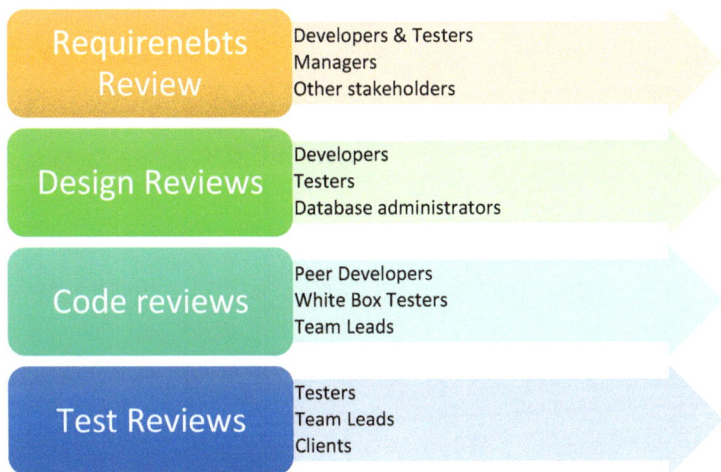

REVIEWERS IN DIFFERENT PHASES

There are multiple roles in a review process and there are certain responsibilities expected in each role. Some of the key roles in the review process are listed below:
- Moderator
- Reviewer / Inspector
- Author / Reviewee
- Manager
- Recorder

Moderator

A moderator trained in the inspection technique conducts the meeting. This role exists only during the meeting. The main responsibility of a moderator is to maintain the decorum and the pace during the review meeting. Sometimes, also known as Review leader, his/ her role is to determine the type of review, approach, and the composition of the review team.

Role:

Session Management
- Ensures that the participants work as a team
- Keeps the meeting objective, professional and friendly
- Avoids needless criticism and keeps the session moving
- Gains final concurrence on problems and rules on those remaining unresolved
- Contributes as a Reviewer
- Ensures that failure to inspect is reported

Concluding the Meeting
- Directs Recorder to review rework items
- Summarizes defects

Reviewer
The reviewer can be a single person or can be a group of people who are reviewing the document. The main responsibility is preparing themselves for the review meeting, i.e. searching for faults and unclear issues while using the allocated time for this task. It is really important to understand the purpose of review and the jargon used and the formats to be used in this work product.

The role also includes checking defects and further improvements in accordance with the business specifications, standards and domain knowledge. The reviewer might have participated in a review as a reviewer and/or as a reviewee.

Role:
Prior to the Meeting
- Inspect the material(s)
- Utilize reference documents and Checklists
- Record problems
- Forwards rework items to Author prior to the meeting (optional)

During the Meeting
- Highlight potential problems.
- Focus on product, NOT producer
- Highlight positive product characteristics

Following the Meeting
- Verify rework as assigned

Author/ Reviewee
The author of the document under review is usually responsible for the investigation of the discovered problems and for carrying out the necessary changes. As the writer of the 'document under review', the author's basic goal should be to learn as much as possible with regard to improving the quality of the document.

The author's task is to illuminate unclear areas and to understand the defects found. The feedbacks and defects should be taken as improvement areas instead of criticism.

Role:
Prior to the Meeting
- Assist their Manager in selecting participants
- Schedule meeting
- Prepare and distribute inspection material

During the Meeting
- Provide resolution date & Provide clarifications
- Not be defensive

Following the Meeting
- Perform rework & Submit rework for verification or re-inspection

Manager
The main responsibility of the manager is to provide resources (time as well as people and process) for review. The manager should allocate proper time for reviews in the project plans.

Role:
Prior to the Meeting
- Allocates time in project schedules
- Assist utilizing reference documents and Checklists
- Determines whether review process objectives

Following the Meeting
- Collate and verify the metrics data

Recorder
The recorder notes down the proceedings in the review meeting. He has to record each defect found and any suggestions or feedback given in the meeting for process improvement. He also keeps track of the suggestions and recommendations made by the reviewers.

Role:
- Makes the Summary & Completes Rework items list
- Reviews rework items at the end of the meeting
- Classify Errors
- Readout/present full list to inspection team at the end of the inspection
- Prepares and share the Minutes of meeting to all review team members.

There can be many other roles in the review process, but we only discussed the most common ones here.

Agile Reviews

Agile methodology is a lot different from other reliable traditional methodologies. In agile, the complete project is divided into small sprints. A sprint is a period of time allocated for a particular phase of a project. Sprints are considered to be complete when the time period expires.

The agile methodology allows agile teams do a little of everything, rather than doing all of one thing at a time. This saves a lot of time, especially in a project where the requirements keep on changing and the deadlines are short.

The reviews become more critical in the agile projects, as one project is broken into small different pieces done by a different team in parallel.

Are we making the product right?
Are we correctly building the solution?

These two are very important questions in Software Development Lifecycle, and failure to answer these questions positively throughout the lifecycle can result in delivering a defective solution. It may cost much more if we fail to verify the solution at each level of construction.

Like any other methodology, reviews help in reducing the cost and effort in the case of agile methodology too. But it is really important to make these reviews part of the project plans to make them align.

In agile projects, there are many challenges in the preparation of documentation, like:

- In traditional software development projects, there are specification documents such as requirement specification, architecture specification, or design specification. In an agile project, there are executable requirements in the form of tests.

- As a part of the agile strategy, you document as late as possible, only before you require them which means the best time to write system requirements will be towards the end of a release. This way you document what you have actually built. The majority of the user and support documentation is pushed to the end of the lifecycle to ensure high quality.

- With the thought of, the higher the number of pages in the document, the higher would be the chances of error, the size of the document is usually very limited, sometimes just the bullet points. It is very difficult to find a complete requirement document to answer the queries.

- The requirements are ever-changing in agile projects. Hence the documentation also keeps on changing. It is often possible to find create Wikis sort of documentation to create different single pages for single topics.

Therefore, getting a hardcore requirement document or test document is difficult in agile projects. The documentation is usually to cater to the situation in hand.

Agile Development Cycle

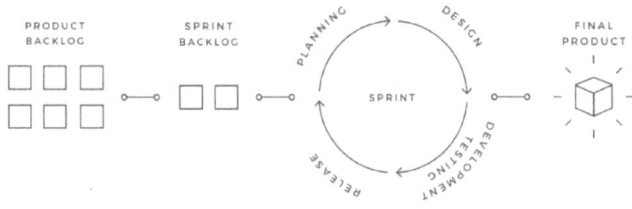

© AltexSoft Inc

An agile project requires a certain number of documents, just like any other methodology. Here we have outlined the most common:

Product Requirements & Design Documents
In the case of agile projects, it is more likely that all requirements will not be identified in the beginning or that they may be improperly defined. In Agile, we

have the luxury of defining the requirements in chunks and thus taking as many Sprints as necessary to get the desired end product in the form of epics, user stories and acceptance criteria. Hence, this makes the review activity a bit difficult.

The review of requirements and design documents may be a complex process as the reviewer will not have the complete requirements in the beginning, but a process to review the epics or the user stories may still help in identifying the missing requirements and improving the quality of requirements.

However, in the case of design documents, it is important to review and get approval on the designs before the development starts. The design is the core document on the basis of which other teams may start their work. Therefore, getting approval on the design and consecutively on any change within the design document is very necessary, especially in the presence of partial requirements.

Testing Documentation
The testing documentation in agile projects usually has Test strategy, Test plan, Test case specifications, and Test execution checklists/ log.

A **test strategy** is a document that describes the software testing approach to achieve testing objectives. This document includes information about team structure and resource needs along with what should be prioritized during testing. A test strategy is usually static as the strategy is defined for the entire development scope in the beginning.

> **Review:** The review of the test strategy document is usually done by the author's manager and the client. The approval on the same is given by the client.

A **test plan** usually consists of one or two pages and describes what should be tested at a given moment. This document should contain:
- The list of features to be tested
- Testing methods
- Timeframes
- Roles and responsibilities
- Entry and Exit criteria

> **Review:** The review of the test plan document is usually done by the test lead, project manager and the approval of the client.

A **test case specifications** document is a set of detailed actions to verify each feature or functionality of a product. Test case specifications are based on the approach outlined in the test plan and are based on the user requirements document.

> **Review:** The review of the test case specifications document is done by all the involved department heads, project managers and the approval of the client.

Test execution checklist/ log is a list of tests that should be run at a particular time. It represents what tests are completed and how many have failed. All points in the test checklists should be defined correctly.

> **Review:** The review of the test checklist document is usually done as a peer review to ensure all possible scenarios are covered.

Maintenance and help guide

This document describes known problems with the system and their solutions. It also represents the dependencies between different parts of the system.

> **Review:** The review of these documents is done by the testing team. They cross-check each line and ensure they are in line with the system.

Process Documentation

Process documentation covers all such documents which are prepared apart from the above-mentioned documents to cater to the activities surrounding the product development. The value of keeping process documentation is to make development more organized and well-planned. Here are common types of process documentation created in a project:

Plans, estimates, and schedules

These documents are usually created before the project starts and can be altered as the product evolves. These documents help in keeping the project on track.

> **Review:** The review of the project plans, estimates, and schedules are done in the initial phase. Throughout the project, the same keeps on getting updated, and a peer review on the same are enough.

Reports and metrics
Reports reflect the time and resources utilization during project processing. They can be generated on a daily, weekly, or monthly basis.

> **Review:** The peer review of the same, before sending out, is a required process to avoid any miscommunication.

Working papers
Working papers are draft documents that exist to record engineers' ideas and thoughts during project implementation, especially in the case of agile projects. Working papers usually contain some information about an engineer's code, sketches, and ideas on how to solve technical issues. While they shouldn't be the major source of information, keeping track of them allows for retrieving highly specific project details if needed.

Review: Such documents do not require any kind of review.

Standards

In a project various standards may be referred, especially recognized within the industry, and certain formats are also considered.

Review: No specific review process is required for the standards as they already are controlled by the publishing organization, but the formats whenever updated, should be reviewed by the key stakeholders.

There are many more documents that may get created in a project. It is really important to get every document reviewed to ensure the quality maintained within the projects.

Review Metrics

After a formal review activity is completed, the metrics data is collected for improving the process. The metrics are created to understand the type of defects found in the reviews. These metrics are usually used to generate training requirements to avoid the similar kind of defects reported in the reviews, in the future.

Recording data about the review process and product quality is a distinguishing characteristic of formal peer reviews. Metrics are collected for quality evaluation of both the inspected object and the inspection process itself. The metrics indicate potential improvements in the process. Entry and exit criteria are enforced on the inspected object to allow for good and stable quality in the inspection process.

These review metrics may give the output in the form of an updated checklist, training needs identification, reorganized formats, reduce rework costs, and improved processes.

While performing a review activity, a number of defects are recorded. These defects are then split into a variety of categories and the metrics are generated. Few metrics related to review defects are mentioned below:

- **Document Review Quality:** The total number of review comments per document is recorded to identify the quality of the document created by the author.

- **Review Effort:** The number of hours spent on reviewing a requirement is calculated and mapped with the complexity of the document.

- **Code Review Effectiveness:** The code review effectiveness is checked via various methods but the most popular is KLOC (Thousand Lines of Code).

- **Rework Effort:** Rework effort is calculated to check how much time is spent in performing a rework.

(Rework Effort/ Creation Effort)* 100

- **Review Efficiency:** The review efficiency is a metric used to reduce the pre-delivery defects in the software. It helps to decrease the probability of defect leakage in subsequent stages of SDLC. The formula for calculating review efficiency is:

Review Efficiency (RE) = Total number of review defects / (Total number of review defects + Total number of testing defects) x 100

Productivity depends on the time you spend on tasks and other in-house activities. To track and increase productivity managers use metrics identifying and prioritizing evolving issues.

Metrics help CEOs and PMs, and other top management members to assess the influence of decisions taken during a development process, sort out priorities and objectives.

Key Guidelines

Now we know what are reviews and how to perform a review. It is really important to keep a few things in mind while performing a review activity. Here are a few checklist points to help.

When to review:
- A new document/software item is prepared
- An Existing Document or Software is modified and is ready for review
- Project receivable is received from the customer
- The review is due as per the Quality Management Plan

Entry criteria for review:
- Has the preceding life cycle activity been concluded?
- Are there any changes to the baseline?
- Are review participants in place and briefed?
- Have all participants received all the review materials and checklists?
- Is the document ready for review?
- Has the review due as per the project plan?

Exit Criteria for review:
- Have all the product elements been inspected?
- Have all checklists been processed?

- Have the inspection results been recorded?
- Have metrics been collected?
- Has the recorder read back the issues?
- Has the producer been given an opportunity to comment, if required?
- Has the author incorporated all the review comments?

Review Guidelines:
While the review is performed, it is important to keep a few things in mind:
- Keep it short (< 30 minutes).
- Don't schedule two in a row.
- Don't review product fragments.
- Use standards to avoid style disagreements.
- Let the coordinator run the meeting and maintain order.
- Use effective formal reviews as a filter prior to testing.
- Conduct formal technical reviews to assess the test strategy and test cases.
- Program reviews/ code reviews are done in the unit testing stage.
- Configuration review or audit is used to ensure that all elements of the software configuration have been properly developed, cataloged, and documented to allow its support during its maintenance phase.

Software Project Reviews

About the Author

Richa Yamini Goel is a Certified Lead Auditor for ISO 9001:2015 and ISO 27001:2013 management standards. She has been associated with multiple certification bodies for external & internal audits around ISO certifications and has also assisted several organizations in ISO implementation and process optimization to improve overall efficiency.

She has worked with multiple IT firms in the field of Software Testing for a decade. She has lead several projects to take organisations through the successful implementation of a number of standards including ISO/IEC 27001 and PCI-DSS and has developed effective security architectures and programmes.

With more than 20 years of experience in distinct verticals of industry, she is currently serving as a Corporate Trainer imparting technical / non-technical trainings and an auditor as well as consultant for ISO certification implementation.

Richa is exceptionally strong in the areas of risk management, information security and business continuity. Her core competencies lie in the space of ISO Standards and Software Testing based programs. She is an author of several articles in IT magazine Digit, TutorialsPoint.com and plenty of books available on Amazon.

Her following books are available on Amazon as well as Google:

- Mulberry Tree and other stories
- Quick Reference Guide for ISO 9001:2015
- Quick Reference Guide for Risk Management

For any queries/ feedback/ suggestions, you can directly contact her at richa@gyanedgeconsulting.com.

Or follow on:

@richayaminigoel
@richa.yamini1
www.gyanedgeconsulting.com

Software Project Reviews

www.ingramcontent.com/pod-product-compliance
Lightning Source LLC
Chambersburg PA
CBHW040234220526
45473CB00001B/240